SEBRING
12-HOUR RACE 1970
PHOTO ARCHIVE

SEBRING 12-HOUR RACE 1970

PHOTO ARCHIVE

Edited by with introduction by
Robert C. Auten

Iconografix
Photo Archive Series

Iconografix
P.O. Box 18433
Minneapolis, Minnesota 55418 USA

Library of Congress Card Number 94-77482

ISBN 1-882256-20-4

94 95 96 97 98 99 00 5 4 3 2 1

Cover and book design by Lou Gordon, Osceola, Wisconsin

Printed in the United States of America

Book trade distribution by Voyageur Press, Inc. (800) 888-9653

PREFACE

The histories of machines and mechanical gadgets are contained in the books, journals, correspondence and personal papers stored in libraries and archives throughout the world. Written in tens of languages, covering thousands of subjects, the stories are recorded in millions of words.

Words are powerful. Yet, the impact of a single image, a photograph or an illustration, often relates more than dozens of pages of text. Fortunately, many of the libraries and archives that house the words also preserve the images.

In the *Photo Archive Series*, Iconografix reproduces photographs and illustrations selected from public and private collections. The images are chosen to tell a story—to capture the character of their subject. Reproduced as found, they are accompanied by the captions made available by the archive.

The Iconografix *Photo Archive Series* is dedicated to young and old alike, the enthusiast, the collector and anyone who, like us, is fascinated by "things" mechanical.

The field in position for the 1970 Sebring 12-Hour Race.

INTRODUCTION

For Alec Ulmann, who had struggled for 20 years to make the 12-Hours of Sebring a success, 1970 was a banner year. A record crowd of 57,500 viewed a race that ended in the closest finish in Sebring history—22.1 seconds separated the first and second place finishers. During the twelve hours, there were a record fifteen lead changes. The winning car, a Ferrari 512S, averaged 107.03 mph and covered a distance of 1,289.6 miles, both new records.

Things had not always gone so well at Sebring. In the early years, Ulmann's AAA sanctioning had the SCCA threatening to run conflicting races the same weekend. Eventually Ulmann's lifetime SCCA membership was cancelled and he was expelled from the organization. Battle lines were drawn on one major issue: amateurism versus professionalism. Ulmann wanted to bring European-styled, factory-backed road racing to the United States. This meant paid drivers. Up until Sebring, road racing in the United States was under the complete control of the SCCA. After the 1955 Sebring race and the subsequent Le Mans disaster of that same year, the AAA decided to withdraw from the sanctioning of automobile racing. USAC took control of oval track racing but was not interested in road racing. This forced Ulmann to start his own organization, the ARCF, and to apply to the F.I.A. for sanctioning. Sebring became the only race in the U.S.A. to receive F.I.A. sanctioning, other than the Indianapolis 500. For eleven years Ulmann carried on alone. In 1959 he even managed to hold a U.S. G.P. Formula 1 race at Sebring. During the intervening years, the SCCA grew up—realizing that there was a place for both the amateur and the professional in U.S. road racing. In 1967, the SCCA became the sanctioning body at Sebring and remained so until 1973 when IMSA took over.

The 1970 Sebring race proved memorable for reasons beyond those expressed above. For Ferrari, the win was its first Manufacturers Championship victory since April 1967. For Mario Andretti, it was a hard won victory. Al Bochrock quoted Andretti in the June 1970 issue of *Road & Track*, "I made more contacts than any race I've ever been in. I hit a dozen of them. Those poor guys spend more time looking backward in their mirrors than looking where they are going. . . It was a balls-out race right from the start. . . I guarantee you I never drove so hard in my life." Twenty four years later, the victory remained as one of his most cherished. Andretti was quoted in *Autoweek* May 16, 1994, "Then there was Sebring (1970), I took over from (Nino) Vaccarella and (Ignazio) Giunti and drove like a man possessed to get what I wanted. Races like that are good to think about over a glass of wine."

Perhaps, however, what overshadowed the events of that weekend was not a new record but a new name in the racing world—Steve McQueen. Driving with a broken foot and paired with playboy driver, Peter Revson, the McQueen-Revson Porsche 908 led the race with ten laps to go. Liberal rules allowed Mario Andretti to switch from his #19 Ferrari 512S to the #21 Ferrari 512S of Giunti and Vaccarella in the last few laps, and kept the McQueen-Revson Porsche out of the victory circle. As it was, theirs was the only other car on the final lap. Perhaps the record crowd in 1970 contained more papparazzi and movie fans than racing enthusiasts, but it can be safely concluded that the presence of Steve McQueen was a major drawing factor.

Aerial view of the front straight and parking facilities at Sebring.

The Gulf-Porsche team lined-up for pre-race inspection.

Ground clearance check.

Gulf-Porsche 917 weighing-in.

The Whitaker-Slotagg-Davidson Volvo 122S at weigh-in. This car completed 25 laps before an oil pressure problem forced its retirement.

Gulf
Gulf
14
FRAM
FILTERS

The Siffert-Redman Gulf-Porsche 917 following inspection.

Gulf-Porsche training car.

RACE DAY
MARCH 21, 1970

Ferrari factory pits before the start of the race.

The Andretti-Merzario Ferrari 512S, one of four entered at Sebring 1970.

One of the two 3-liter Ferrari 312 prototypes entered by Luigi Chinetti.

Paul Pettey and Roy Hallquist drove this 5.0 liter Mustang to 27th place.

The Stommelen-Galli Alfa T33-3.

The Dr. V. P. Collins-Larry Wilson Mustang prior to the start. Collins and Wilson finished the race 24th overall.

Rear view of the pits.

Two North American Racing Team (N.A.R.T.) mechanics at their trailer behind the pits.

The 16th place finisher, the Robert Mitchell-Charlie Kemp Chevrolet Camaro.

The Vince Gimondo-Chuck Dietrich Chevy Camaro placed 14th overall and first in its category.

BMC roadster driven by Reggie Smith, Donley, and Butari.

The Ben Scott, Lowell Lanier, Dave Houser MGB, and the Smith BMC in their pit.

On the track, prior to the start of the race.

John Wyer Gulf-Porsche 917 prior to the race.

Lining up the cars prior to the start.

Alec Ulmann, creator of the 12-Hours of Sebring, readies the green flag.

Three-time Sebring winner Phil Hill paced the 68 starters past the line in Sebrings first rolling start.

The race is on! Pole sitter Mario Andretti leads the pack.

Dick Lang and co-driver Tony de Lorenzo drove the Owens-Corning Fiberglass Corvette to 10th place overall and a 1st place finish in the GT class.

Dick Lang at the wheel and the Stommelen-Galli Alfa Romeo T33-3.

The Corvette Stingray driven by Bill Schumacher and Bill Petree was retired after 34 laps.

The team of Bob Grossman and Don Yenko drove their 7-liter Chevrolet Camaro to a 17th place finish.

The Jo Siffert-Brian Redman Porsche 917 lost the pole position to Mario Andretti's Ferrari 512S by 0.57 seconds.

Front end damage forced the withdrawal of the Siffert-Redman Porsche 917 after 211 laps.

The Rodriguez-Kinnuen 917 shadowed by the Ickx-Shetty Ferrari 512S. Driven by Pedro Rodriguez, Leo Kinnuen, and—late in the race—Jo Siffert, it placed 4th overall.

The Rodriguez-Kinnuen 917 overtaking the Gregg-Harrison Porsche 911.

The Vic Elford-Kurt Ahrens factory Porsche 917 was withdrawn following an accident on its 62nd lap.

Rear view of the Elford-Ahrens Porsche 917.

Officials tower on the backside of the track.

The Elford-Ahrens 917 and the Chinetti-Adamowicz Ferrari 312 P. Both cars retired early: the Ferrari, due to overheating, after 56 laps; the Porsche, as stated, due to an accident.

Paul Pettey and Roy Hallquist drove this 5.0 liter Ford Mustang to 27th place.

The Pettey-Hallquist Mustang leaves the pits.

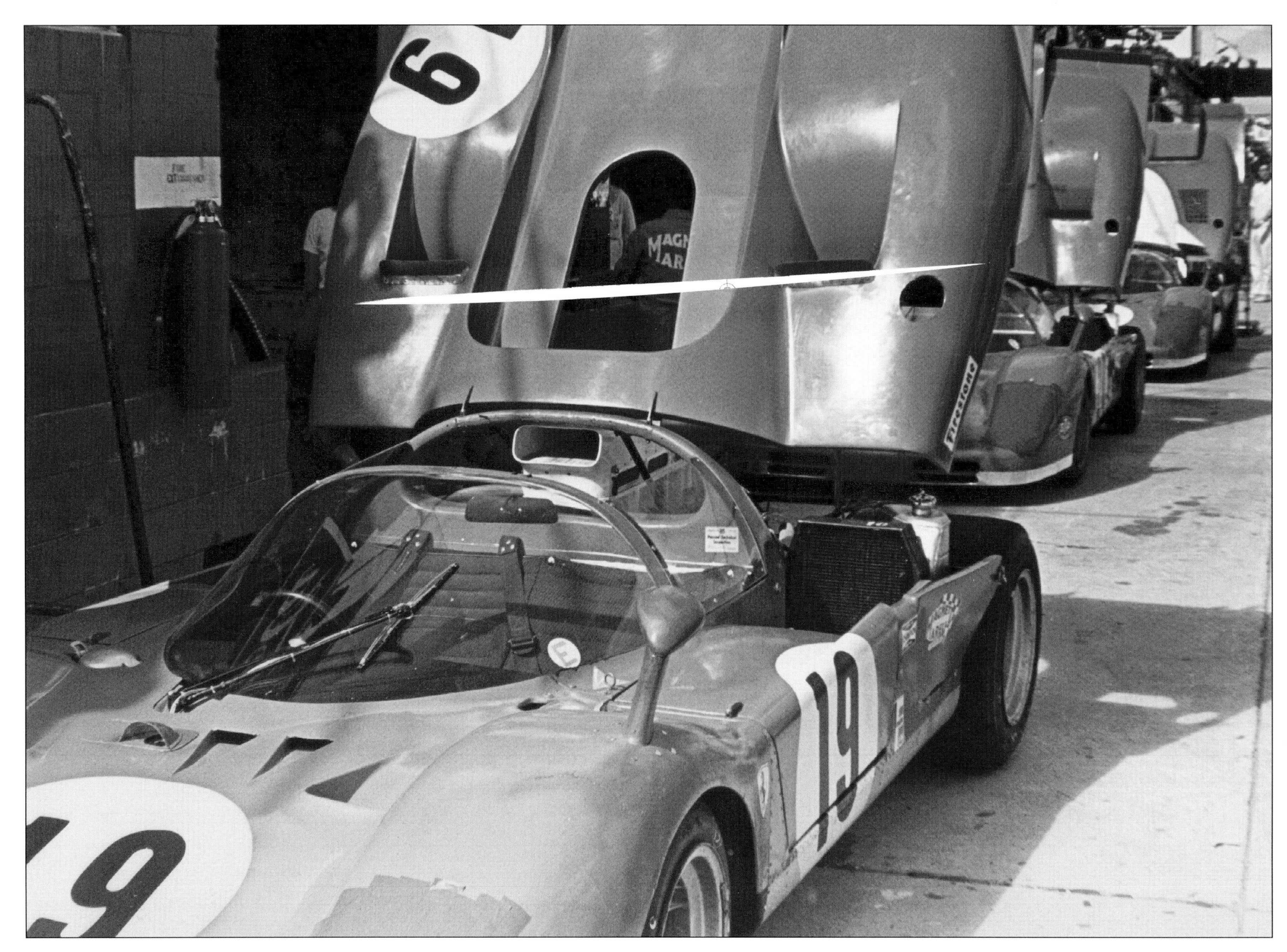

The Mario Andretti-Arturo Merzario factory Ferrari 512S in its pit.

19
19
19
STP

The Andretti-Merzario Ferrari 512S was retired after 227 laps with gearbox trouble.

The Jacky Ickx-Peter Shetty Ferrari 512S, one of three factory 512Ss entered at Sebring 1970.

The winning Ferrari 512S, driven by Ignazio Giunti, Nino Vaccarella, and Mario Andretti.

Firestone
21
21
STP

Ignazio Giunti takes the wheel.

The Giunti-Vacarella-Andretti 512S completed the race at a record average speed of 107.290 mph.

The winning Ferrari under darkness of night.

The Mike Parkes-Chuck Parsons Ferrari 312P.

Chuck Parsons takes the wheel of the N.A.R.T. 312P. This car finished 6th overall.

The Parkes-Parsons' pit crew at work.

The second of two 312Ps entered by Chinetti and driven by Luigi Chinetti, Jr. and Tony Andamowicz.

25

24
GOOD/YEAR

Ronnie Bucknum and Sam Posey piloted this Chinetti Ferrari 512S.

Sam Posey brings his Ferrari 512S into the pits.

The Posey-Bucknum 512S retired after 92 laps with a fuel pump problem.

The Harley Cluxten-Dr. Wilbur Picket Ferrari 512S withdrew prior to the race.

The Ford GT 40, driven by Ray Heppenstall, F. Grant, and Buzz Marcus, was retired after 117 laps due to a broken rear axle.

The Heppenstall-Grant-Marcus GT 40, the Chinetti-Adamowicz Ferrari 312P, and the Collins-Wilson Mustang vie for position.

The Piers Forester-Andrew Hedges Ford GT 40 completed 22 laps and withdrew with engine problems.

The Piers Courage-Andrea De Adamich Alfa Romeo T33-3 finished eighth overall.

The Rolf Stommelen-Nanni Galli Alfa T33-3.

The Stommelen-Galli Alfa T33-3 finished ninth overall.

The Stommelen-Galli Alfa T33-3 leading the Elliott-Gwyne Camaro (22nd overall) through a corner.

The Stommelen-Galli Alfa T33-3 leading the Kennedy-Samm Lancia Fulvia through a corner.

The Masten Gregory-Toine Hezemans Alfa Romeo T33-3 finished third overall, one lap back.

The Gregory-Hezemans T33-3 leads the Kennedy-Samm Lancia Fulvia, and the Morehead-Vega American Motors AMX (DNF).

33
BOSCH
33

The Henri Pescarolo-Johnny Servoz-Gavin Matra-Simca 650.

34
34
MATRA
SIMCA
elf

François Cevert at the wheel of the Gurney-Cevert Matra-Simca 650.

Dan Gurney in command of the Gurney-Cevert Matra-Simca 650.

35
35

The two Matra-Simca 650s, No. 34 and 35, finished 5th and 12th, respectively.

The McClain-Goo-Floyd Chevrolet Camaro withdrew after 47 laps with engine failure.

Porsche 908 in the pits with driver (either Gerard Larrousse or Gerhard Koch).

Porsche 908 driven to 7th overall by Hans-Dieter Dechent and Tony Dean.

Steve McQueen's Porsche 908 co-driven by Peter Revson.

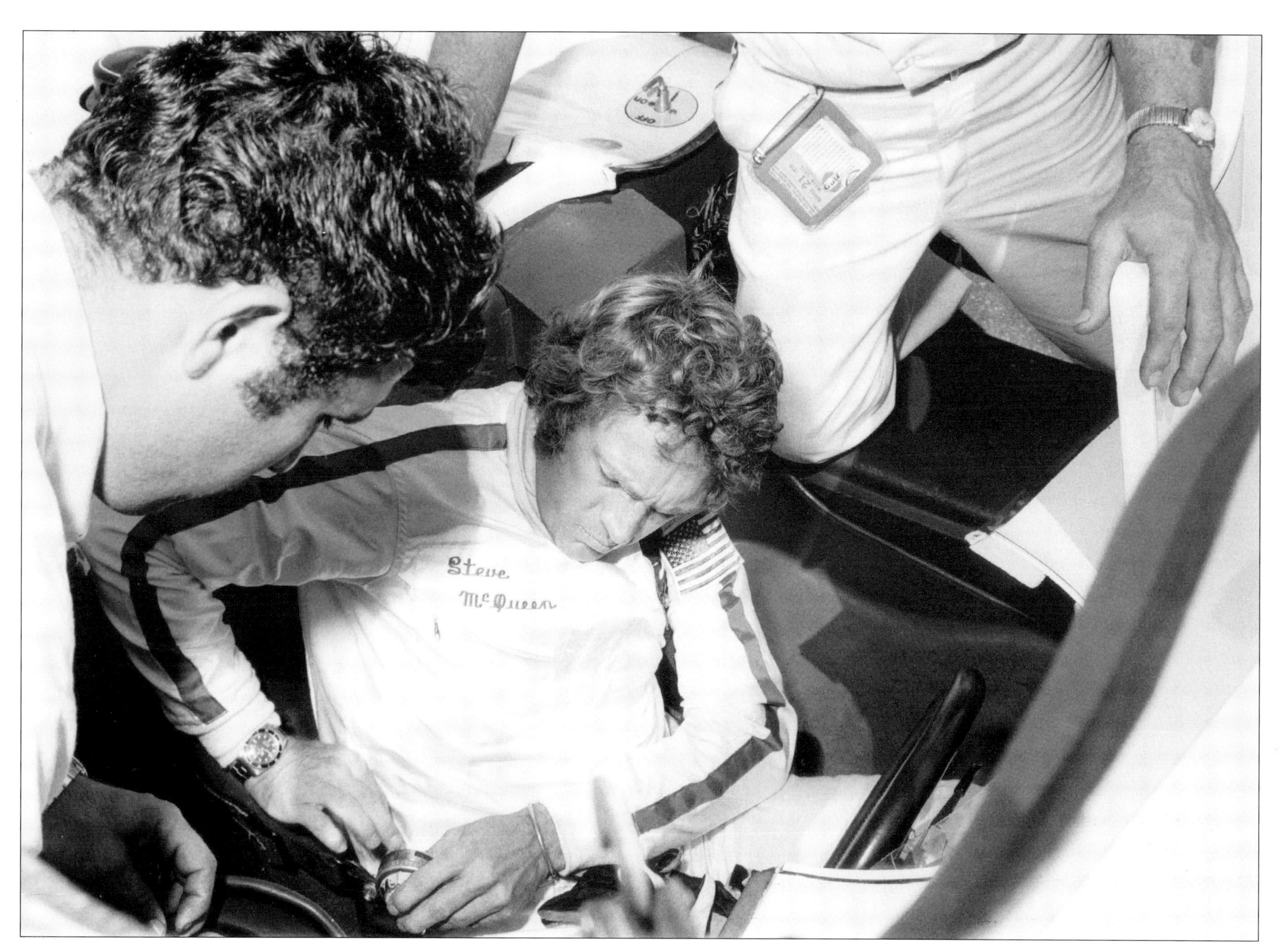

Steve McQueen settles into his Porsche 908.

48
PORSCHE

The McQueen-Revson car in for fuel.

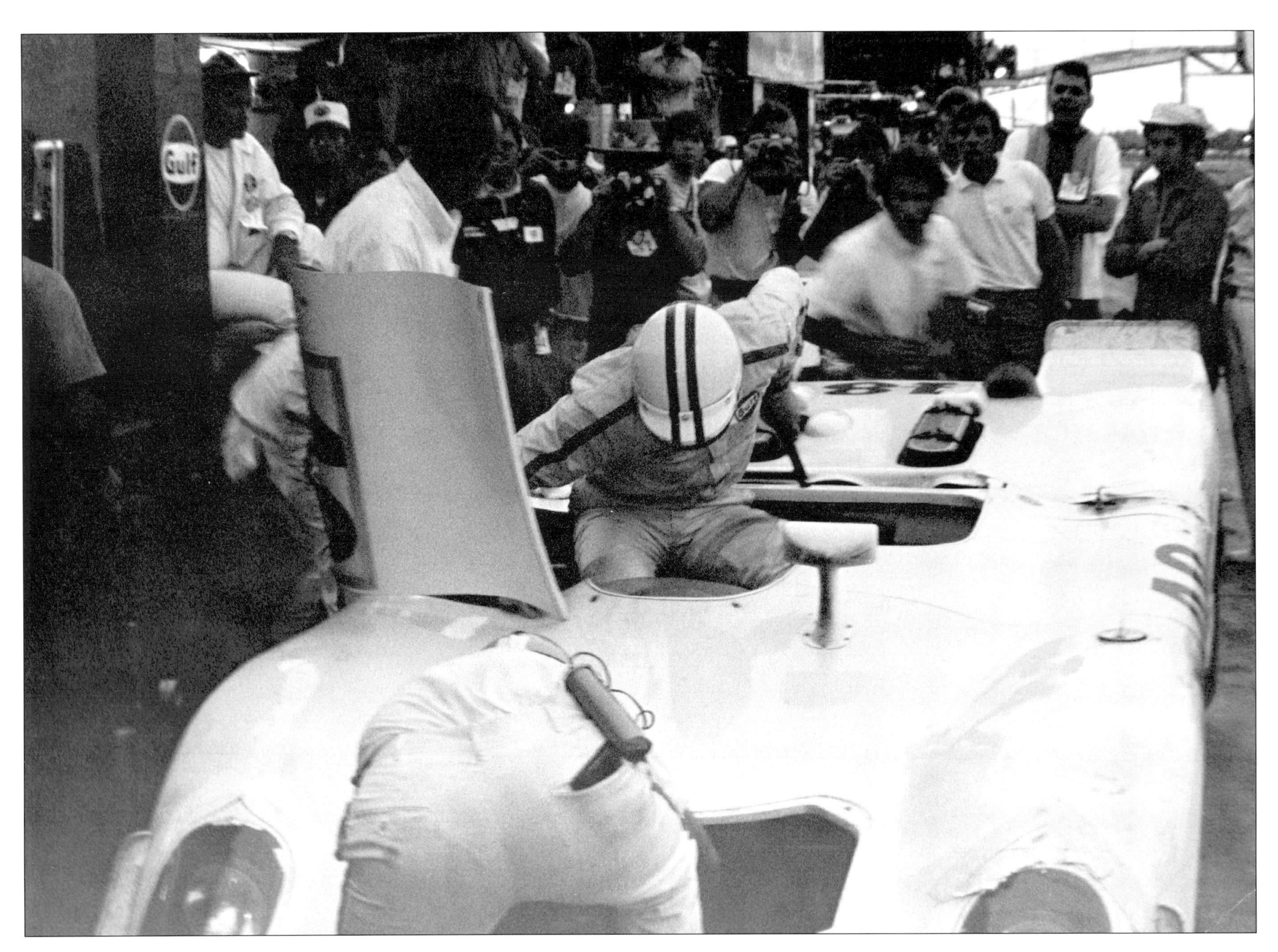

The McQueen-Revson Porsche 908.

Porsche 910 driven by Josef Greger and Andreas Schmalback was withdrawn after 22 laps with engine problems.

Porsche 911 driven by Peter Gregg and Peter Harrison to 12th place overall..

53

Merle Brennan at the wheel of the Brennan-Logan Blackburn MGB.

The Brennan-Blackburn MGB was withdrawn after 84 laps with a broken oil line.

The Waldron Motors MGB driven by John Belperche, Jim Gammon, and Ray Mummery to 25th place overall.

The Smith-Donley-Buttari BMC in its pit.

Chevron B16 driven by Brian Robinson and Hugh Kleinpeter.

Lotus 47 driven by Jim Bandy and Fred Stevenson was withdrawn after 48 laps with rear hub problems.

Alfa Romeo Spyder driven by Paul Spruell and Harry Theodorocopulos was withdrawn after 75 laps with electrical problems.

Porsche 911T driven by Bruce Jennings and Bob Tullius was withdrawn after 78 laps with engine problems.

The Kennedy-Samm Lancia Fulvia followed by the Pescarolo-Servoz-Gavin Matra-Simca 650, and the Andretti-Merzario Ferrari 512S.

Chevrolet Camaro driven by Jim Corwin and Donna Mae Mims to 21st place overall.

Crosley Hot-Shot track runabout.

Sebring at night.

THE PERSONALITIES

Alec Ulmann and the chief pit and paddock marshall just before the start of the race.

Steve McQueen, film star and race driver, relaxes before the race.

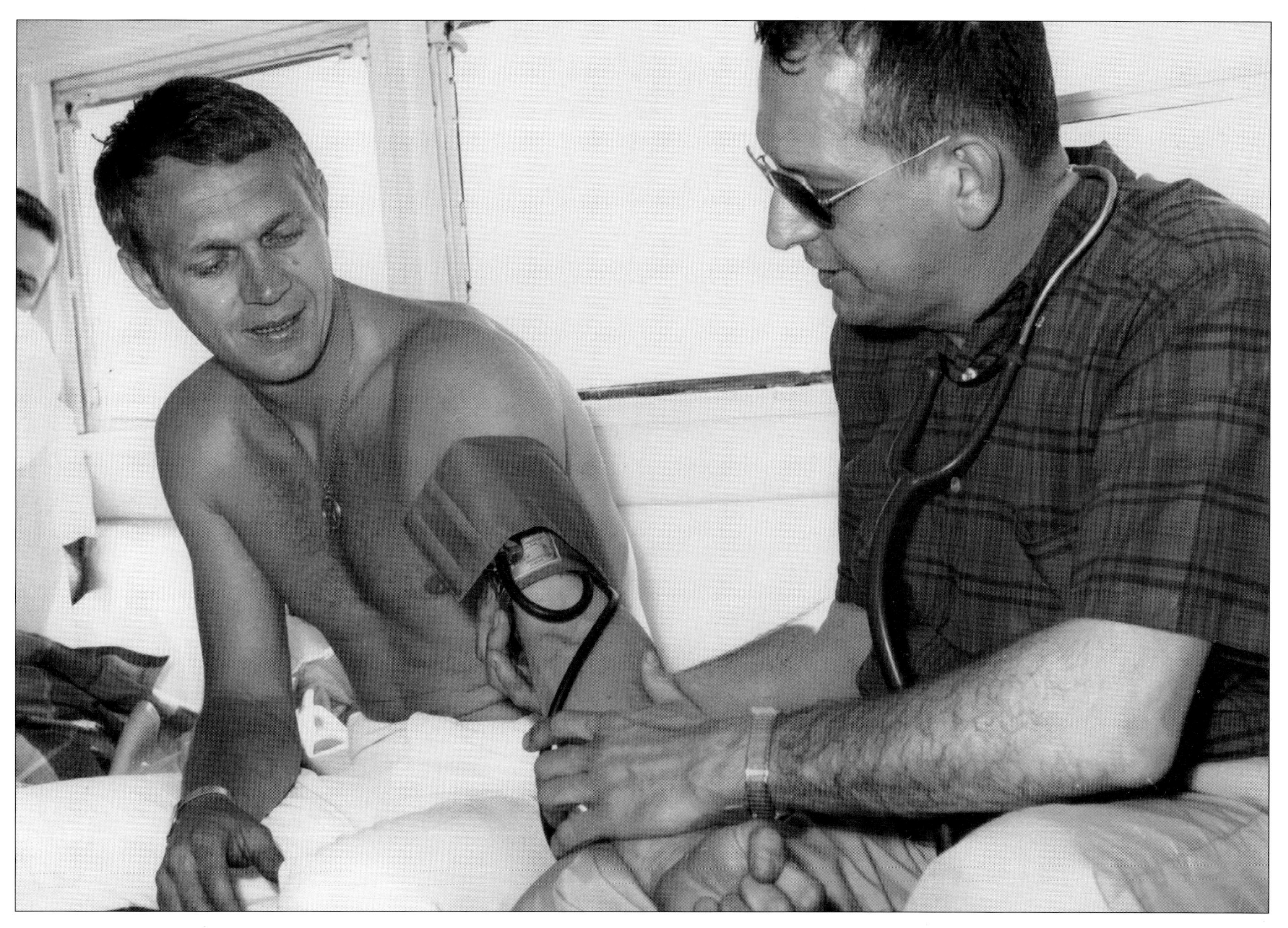

Steve McQueen gets a final medical check-up from Dr. Wallace, Sebring's head medical officer. McQueen drove the race with a broken foot.

Jo Siffert and Steve McQueen prior to the race.

Andrew Furgeson, team manager,
Steve McQueen, and Peter Revson.

Fiat Abarth Spyder drivers, Anatoly Arutunoff and Bill Pryor.

Andrew Hedges, the co-driver of Ford GT 40 No. 30.

François Cevert, the co-driver of Matra-Simca 650 No. 35.

Gulf Porsche driver awaiting late driver change.

Masten Gregory in the cockpit of Alfa Romeo T33-3 No. 33.

Mario Andretti chatting with reporters prior to the race.

Perdro Rodriquez, a co-driver of the John Wyer Gulf-Porsche 917 No. 15.

Janet Guthrie, the co-driver of Austin Healey Sprite No. 73 which finished 19th overall and first in its class.

Brian Redman, the co-driver of Gulf Porsche 917 No. 14.

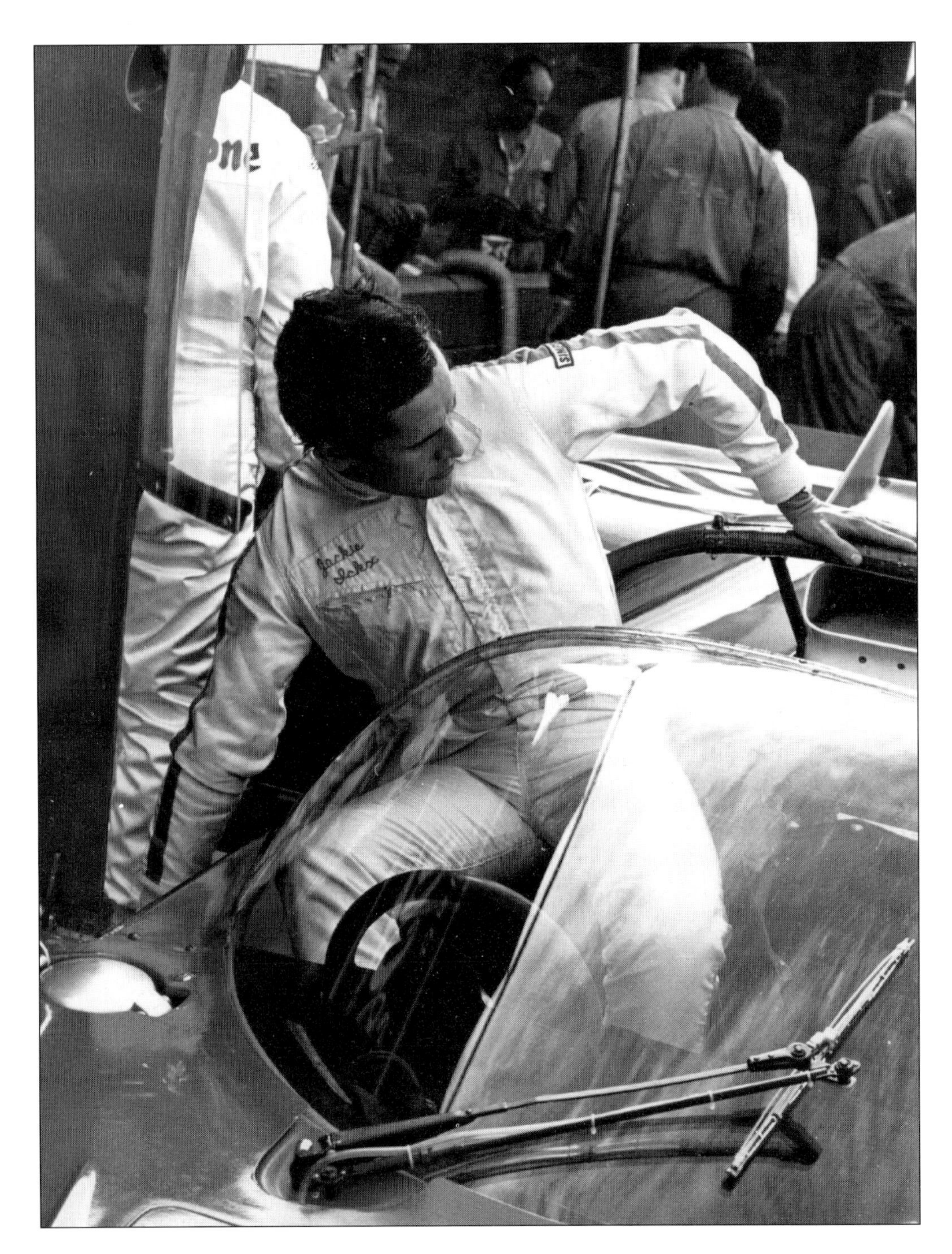

Jacky Ickx, the co-driver of Ferrari 512S No. 20.

Jackie
Ickx
Gulf

Jo Siffert, co-driver of Gulf Porsche 917 No. 15.

Vic Elford, the co-driver of Gulf Porsche 917 No. 16.

Leo Kinnunen, a co-driver of Gulf Porsche 917 No. 15.

Gulf

Chuck Parsons, the co-driver of N.A.R.T. Ferrari 312P No. 22.

Don Yenko, the co-driver of Chevrolet Camaro No. 9.

Henri Perscarolo, the co-driver of Matra-Simca 650 No. 34.

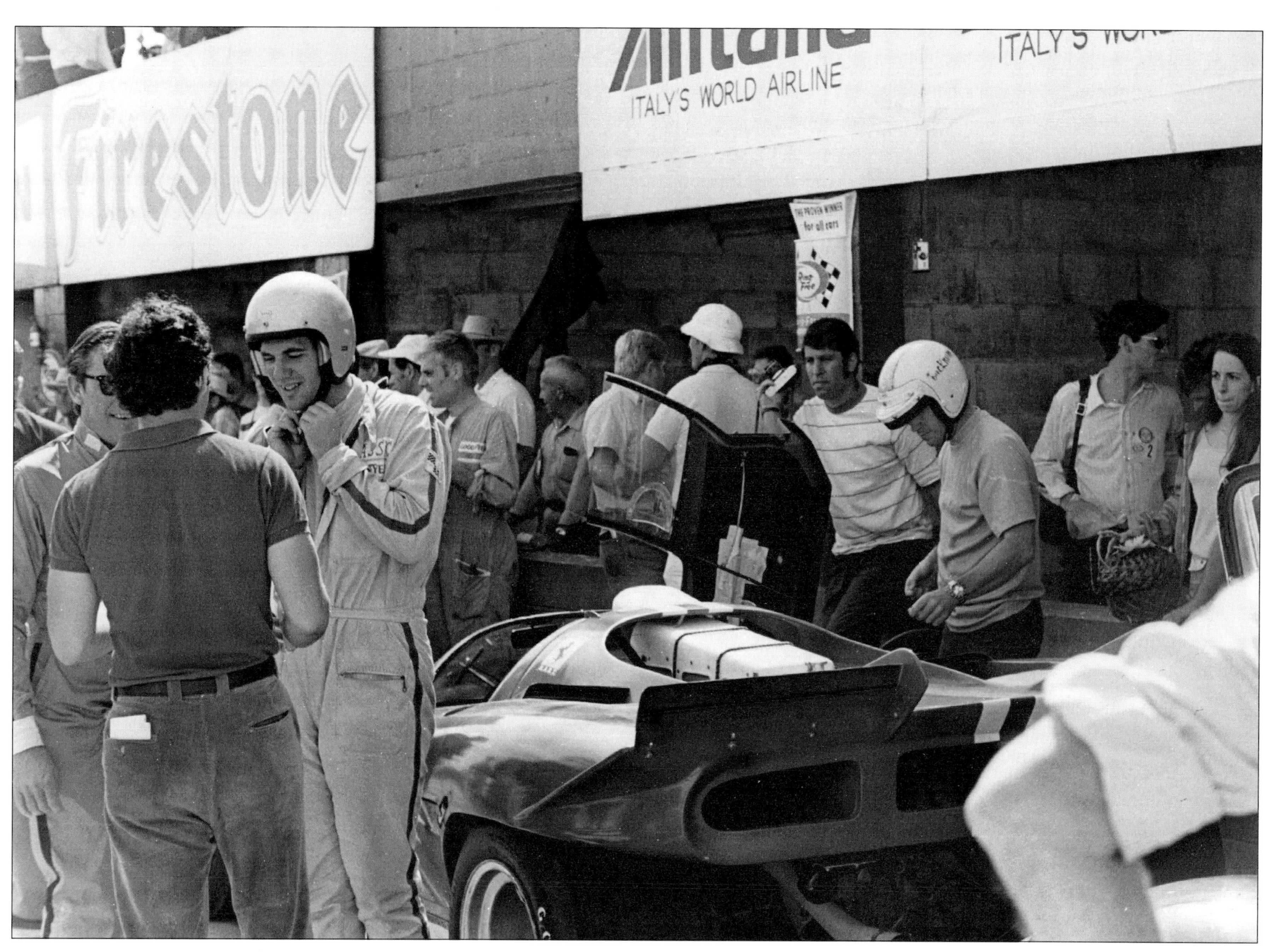

Sam Posey (helmeted, at right) and co-driver Ronnie Bucknum.

Peter Schetty relaxes before the race.

Mike Parkes, the co-driver of the Chinetti Ferrari 312 prototype No. 22 .

Mario Andretti in the victory circle.

THE 64 SEBRING COMPETITORS

Pos.	No.	Manufacturer/Type	cc	Team	Laps
1	21	Ferrari 512S	4994	Andretti—Vaccarella—Giunti	248
2	48	Porsche 908	2997	McQueen—Revson	248
3	33	Alfa Romeo T33-3	2993	Gregory—Hezemans	247
4	15	Gulf Porsche 917	4494	Rodriguez—Kinnunen—Siffert	244
5	34	Matra-Simca 650	2999	Pescarolo—Servoz-Gavin	242
6	22	Ferrari 312P	3000	Parkes—Parsons	240
7	46	Porsche 908	2997	Larrousse—Koch—Attwood	236
8	31	Alfa Romeo T33-3	2993	Courage—de Adamich	231
9	32	Alfa Romeo T33-3	2993	Stommelen—Galli	230
10	1	Chevrolet Corvette	7000	de Lorenzo—Lang	219
11	3	Chevrolet Corvette	7000	Johnson—Johnson—Greendyke	214
12	35	Matra-Simca 650	2999	Gurney—Cevert	213
13	53	Porsche 917	1991	Gregg—Harrison	205
14	40	Chevrolet Camaro	5000	Gimondo—Dietrich	203
15	52	Porsche 911T	1991	Duval—Balley	201
16	39	Chevrolet Camaro	5000	Mitchell—Kemp	191
17	9	Chevrolet Camaro	7000	Grossman—Yenko	189
18	43	Chevrolet Camaro	4997	Tremblay—McDill	187
19	73	Healey Sprite	1300	Smith—Guthrie	187
20	2	Chevrolet Corvette	7000	Thompson—Mahler	187
21	91	Chevrolet Camaro	4956	Corwin—Mims	184
22	92	Chevrolet Camaro	4956	Elliott-Gwynne	182
23	50	Porsche 908	1991	Rahal—Wise—Frank	181
24	37	Ford Mustang	4655	Collins—Wilson	175
25	57	MGB	1798	Belperche—Gammon—Mummery	175
26	58	MGB	1798	Scott—Lainer—Houser	169
27	18	Ford Mustang	5000	Pettey—Halquist	162
28	80	Lancia Fulvia	1300	Kennedy—Samm	157

DID NOT FINISH

No.	Manufacturer/Type	Team	Laps
27	Lola Chevrolet T70	Da'udy—Hallwood	1
79	Alfa Romeo GTV	Talor—Dyler—Drolet	2
59	BMC	Smith—Donley—Buttari	2
60	MGB	Kilpatrick—Goodrich	7
7	American AMX	Morehead-Vega	2
86	Volvo 122S	Theall—Polimeni	8
36	Ford Mustang	Cuomo—Gimbel—Lisberg	16
49	Porsche 910	Greger—Schmalback	22
30	Ford GT 40	Forester—Hedges	22
87	Volvo 122S	Whitaker—Slottag—Davidson	25
17	Porsche 917	Herrmann—Lins	28
45	Porsche 908	Larrousse—Koch	31
5	Chevrolet Corvette	Schumacher—Petree	34
74	Porsche	Meaney—Bean	36
62	Chevron B 16	Robinson—Kleinpeter	38
41	Chevrolet Camaro	Bock—Dent	40
88	Fiat 124	Fleming—Johnson—Bowers	47
38	Chevrolet Camaro	McLain—Boo—Floyd	47
67	Lotus 47	Bandy—Stevenson	48
23	Ferrari 312P	Chinetti—Adamowicz	56
16	Porsche 917	Ahrens—Elford	61
61	Chevron B 16	J. Baker—C. Baker—Richards—Rinzler	70
69	Alfa Romeo	Spruell—Theodoracopulos	75
77	Porsche 911T	Jennings—Tullius	78
82	Lancia Fulvia	Clark—Marsula	78
55	MGB	Brennan—Blackburn	84
24	Ferrari 512S	Bucknum—Posey—Everett	92
26	Lola Chevrolet T70	de Ormes—Young	114
29	Ford GT 40	Grant—Heppenstall—Marcus	117
56	MG Midget	Woodner—O'Connor	122
20	Ferrari 512S	Ickx—Schetty	151
4	Chevrolet Corvette	Greenwood—Barker	159
8	Chevrolet Corvette	Heinz—Costanzo	191
14	Gulf Porsche 917	Siffert—Redman	211
19	Ferrari 512S	Andretti—Merzario	227
47	Porsche 908	Laine—van Lennep	229

The photographs reproduced in this book are a part of the Alec Ulmann estate. They are typical of the type of automobilia offered for sale by Automobilia International. Enquiries from buyers and sellers of automobilia are invited.

Automobilia International
PO Box 606
Peapac, New Jersey 07977 USA
(908) 469-9666

The Iconografix Photo Archive Series includes:

JOHN DEERE MODEL D Photo Archive	ISBN 1-882256-00-X
JOHN DEERE MODEL A Photo Archive	ISBN 1-882256-12-3
JOHN DEERE MODEL B Photo Archive	ISBN 1-882256-01-8
JOHN DEERE 30 SERIES Photo Archive	ISBN 1-882256-13-1
FARMALL REGULAR Photo Archive	ISBN 1-882256-14-X
FARMALL F-SERIES Photo Archive	ISBN 1-882256-02-6
FARMALL MODEL H Photo Archive	ISBN 1-882256-03-4
FARMALL MODEL M Photo Archive	ISBN 1-882256-15-8
CATERPILLAR THIRTY Photo Archive	ISBN 1-882256-04-2
CATERPILLAR SIXTY Photo Archive	ISBN 1-882256-05-0
CATERPILLAR MILITARY TRACTORS VOLUME 1 Photo Archive	ISBN 1-882256-16-6
CATERPILLAR MILITARY TRACTORS VOLUME 2 Photo Archive	ISBN 1-882256-17-4
TWIN CITY TRACTOR Photo Archive	ISBN 1-882256-06-9
MINNEAPOLIS-MOLINE U-SERIES Photo Archive	ISBN 1-882256-07-7
HART-PARR Photo Archive	ISBN 1-882256-08-5
OLIVER TRACTORS Photo Archive	ISBN 1-882256-09-3
HOLT TRACTORS Photo Archive	ISBN 1-882256-10-7
RUSSELL GRADERS Photo Archive	ISBN 1-882256-11-5
MACK MODEL AB Photo Archive	ISBN 1-882256-18-2
MACK MODEL B, 1953-66 Photo Archive	ISBN 1-882256-19-0
LE MANS 1950: THE BRIGGS CUNNINGHAM CAMPAIGN Photo Archive	ISBN 1-882256-21-2
SEBRING 12-HOUR RACE 1970 Photo Archive	ISBN 1-882256-20-4
IMPERIAL 1955-1963 Photo Archive	ISBN 1-882256-22-0
IMPERIAL 1964-1968 Photo Archive Available Early 1995	ISBN 1-882256-23-9
STUDEBAKER 1926-1938 Photo Archive	ISBN 1-882256-24-7
STUDEBAKER 1939-1958 Photo Archive	ISBN 1-882256-25-5
GORDON BENNETT CUP RACE 1905 Postcard Archive	ISBN 1-882256-26-3
AMERICAN SERVICE STATIONS Photo Archive	ISBN 1-882256-27-1
MACK FC, FCSW, NW1936-1947 Photo Archive	ISBN 1-882256-28-X
MACK EB, EC, ED, EE, EF, EG & DE 1936-1951 Photo Archive	ISBN 1-882256-29-8
INTERNATIONAL TD CRAWLERS Photo Archive	ISBN 1-882256-30-1
FARMALL EXPERIMENTAL TRACTORS Photo Archive	ISBN 1-882256-31-X
CASE TRACTORS Photo Archive	ISBN 1-882256-32-8
FORDSON 1917-1928 Photo Archive	ISBN 1-882256-33-6

The Iconografix Photo Archive Series is available from direct mail specialty book dealers and bookstores throughout the world, or can be ordered from the publisher.

For information write to:

Iconografix
P.O. Box 609
Osceola, Wisconsin 54020 USA

Telephone: (715) 294-2792
(800) 289-3504 (USA and Canada)
Fax: (715) 294-3414